Alexander Fleming

Pioneer with Antibiotics

Beverley Birch

**BLACKBIRCH™
PRESS**

THOMSON

GALE

Detroit • New York • San Diego • San Francisco • Cleveland
New Haven, Conn. • Waterville, Maine • London • Munich

THOMSON

GALE

Photo Credits: Bridgeman Art Library: 11, 12 (below), 15, 16; Edinburgh Photographic Library: 12 (top); Mary Evans Picture Library: 19, 21 (top), 28, 29; Nick Birch (Exley Publications Picture Library): 21 (below), 23, 24, 25 (right), 32 (both), 34, 35 (all), 38, 42 (top), 43, 47 (top), 49 (top), 53, 56–57 (all); Popperfoto: 7, 40, 59; Robert Hunt Picture Library: 31, 48, 54; Science Photo Library: 9 (both), 25 (left), 39, 42 (below); Sir William Dunn School of Pathology: 44, 45, 46–47, 49 (below), 50, 51; St. Mary's Hospital Medical School: Cover, 18, 20, 26, 27, 33, 37, 55; The Scottish Regimental Trust: 17; Time Inc.: 5.

LIBRARY OF CONGRESS CATALOGING-IN-PUBLICATION DATA

Birch, Beverley.
  Alexander Fleming / by Beverley Birch.
    p. cm. — (Giants of Science)
Summary: Recounts the life story of Alexander Fleming, his study of medicine and bacteriology, and his discovery of penicillin.
Includes bibliographical references.
  ISBN 1-56711-656-6 (hardback : alk. paper)
  1. Fleming, Alexander, 1881–1955—Juvenile literature. 2. Bacteriologists—Great Britain—Biography—Juvenile literature. 3. Penicillin—History—Juvenile literature. [1. Fleming, Alexander, 1881-1955. 2. Scientists. 3. Penicillin—History.] I. Fleming, Alexander, 1881–1955. II. Title. III. Series.
  QR31.F5 B55 2002
  616'.014'092—dc21   2002003242

Printed in China
10 9 8 7 6 5 4 3 2 1

Contents

FIFTEEN CENTS   MAY 15, 1944

TIME

THE WEEKLY NEWSMAGAZINE

DR. ALEXANDER FLEMING
His penicillin will save more lives than war can spend.
(Medicine)

VOLUME XLIII   NUMBER 20

Introduction:
The Case of Harry Lambert

The man was plainly dying. For more than six weeks, doctors had fought to save him. But now his temperature soared. He tossed and mumbled in the throes of the high fever, wracked by agonizing spasms and uncontrollable hiccups. He also drifted in and out of a coma. To Alexander Fleming, standing by his bedside, there was no doubt that Harry Lambert had little time to live.

Fleming, like the other doctors at St. Mary's Hospital in London, knew what was wrong with Lambert. Diseases like his were usually caused by a microbe—a microscopic living organism—that invaded the person's body and weakened or poisoned it, sometimes to the point of death. The doctors had tried to treat Lambert with the only drugs they had. He had simply become much worse.

As Lambert's illness progressed, Fleming had been trying to identify the microbe that had caused it. Without knowing which microbe it was, the doctors could not hope to save Lambert's life. And so in his laboratory Fleming had worked on determinedly.

At first Lambert had seemed to be suffering from a kind of influenza. But as his condition deteriorated, all the signs of the terrible disease known as meningitis began to appear. And meningitis, an infection of the waterproof sheath of membranes that surround the brain and spinal cord, was frequently fatal.

Using a hollow needle and a syringe, Fleming drew off some of the watery fluid that surrounded Lambert's spinal cord. If Lambert had meningitis, then the microbe that caused the disease would be present in that spinal fluid.

Fleming examined the fluid, and at last he found the microbe. It was one of the round microbes, which grow in chains, called a streptococcus—virulent, fast-spreading, and often the cause of killer diseases.

Doctors had been having some successes killing streptococci with new chemical drugs called

Opposite: A 1944 magazine cover praised Alexander Fleming's discovery of penicillin.

5

sulfonamides. But the doctors had used these drugs on Harry Lambert, and he was still dying.

One Last Chance

There was one last thing to try. Fleming telephoned the only person who could help him: Professor Howard Florey. Florey, in Oxford, headed a team of scientists who had developed a new drug. It was made from a substance that Fleming had discovered nearly fourteen years before, and named penicillin.

Florey's response was swift. He sent Fleming all the penicillin they had in Oxford. It was a precious supply: there would be none left after Fleming used this up, not until the Oxford team could make some more. And that was a slow and laborious process that would take months.

Florey gave Fleming exact instructions for using the substance. It was important to follow these instructions, because penicillin did not stay in the body very long, and it was vital to maintain the supply long enough for it to do its work.

Harry Lambert would not be the first person to receive penicillin injections—Florey and his team had been experimenting with human patients for some time. On August 6, 1942, Lambert, however, would become the first person to be treated with penicillin by the man who had discovered it.

Watching and Waiting

Fleming gave the first injection. Three hours later, he gave another, and another three hours later. He continued the injections throughout the night, while nurses and other doctors watched. And as they watched, Harry Lambert became calmer. The hiccups disappeared, and finally he slept. Only twenty-four hours after the first injection, doctors—almost in disbelief—found that Harry Lambert's temperature had fallen to normal for the first time in more than six weeks.

Howard Florey had made it clear that the penicillin injections would have to continue for seven more days. On August 13, however, Alexander

Fleming became anxious. Harry Lambert had stopped getting better—his temperature was rising, and the delirious ramblings had begun again.

What could Fleming do now? He had found the streptococcus in the watery fluid around Lambert's spinal cord. Had the penicillin failed to pass from the man's blood into his spinal fluid, to kill the streptococci there?

Again Fleming drew off some of Lambert's spinal fluid into a syringe, and took it quickly to his laboratory. He put the fluid under a microscope and checked for signs of penicillin. There were none. If the streptococci had not been killed in the spinal fluid, they were multiplying and spreading, and there was no hope for Harry Lambert.

A Life or Death Gamble

Again Fleming telephoned Florey in Oxford. Florey's team of scientists and doctors had been

Fleming worked in a laboratory at St. Mary's Hospital.

testing penicillin for many months. Fleming asked, had they ever injected it directly into someone's spinal canal, the tunnel in the backbone that held the spinal cord? Florey replied that they had not.

Fleming considered whether to try such an injection on Lambert. On one hand, it might save his life. On the other hand, there was a chance it would so badly shock his body that the injection itself would kill him.

Fleming did not know that even as he made up his mind to take the risk and give Lambert the injection, Howard Florey was also investigating the problem—he was injecting penicillin into the spinal canal of a rabbit. Nor did Fleming know that the rabbit immediately died.

In London, Fleming carefully inserted the needle between two sections of Lambert's spine, into the spinal canal. The penicillin shot straight into the spinal fluid. Fleming waited. Would Lambert start shivering violently? Would his temperature soar? Would he vomit, have a fit, or show any of the signs of severe shock at something that had devastated his body? Might he even die?

As it turned out, none of these things happened. Before Fleming's eyes, a transformation began. First Lambert's temperature dropped. His delirious ramblings stopped and he became clearheaded. All signs of fever and inflammation faded. His appetite returned. As the days went by, he received more injections into his spinal canal. He simply got better and better.

According to all prior medical experience of the time, Harry Lambert should have been dead. Instead, after only a month of recovery time, he walked out of the hospital completely cured.

The Miracle Drug

Harry Lambert owed his life not only to Fleming, but also to the men and women in Oxford—the team of scientists led by Howard Florey. They had taken Fleming's penicillin and purified and refined it. They had worked out by months of trial and error how to keep it in a person's body long enough for it to do its

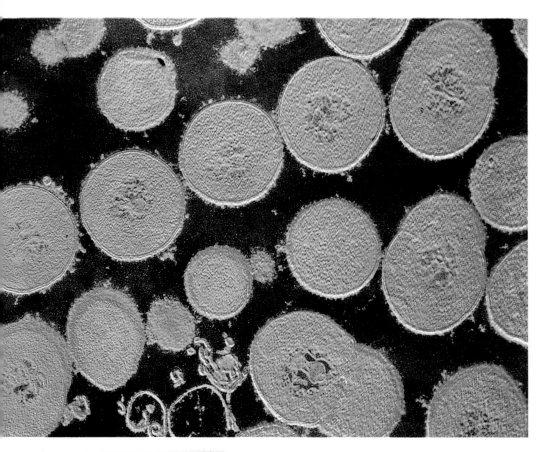

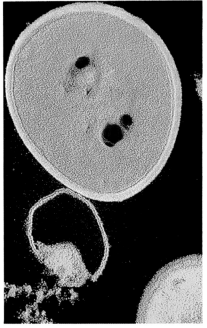

Top: This photograph shows streptococci, the kind of microbes that infected Harry Lambert, magnified more than 25,000 times.

Bottom: This photograph shows Staphylococcus aureus, magnified 60,000 times. At the top is a normal bacterium. Below it is a bacterium that is being dissolved by an antibiotic. The antibiotic destroys the outer membrane—which appears blue around the normal bacterium—and the contents of the bacterium spill out.

lifesaving work. They had seen its power, and were probably the only people who fully understood that this drug, created from the raw juice of a mold, had the potential to save untold millions of lives. Only in those August weeks of 1942 did Fleming himself see the miraculous effect of his own discovery.

Pasteur's Revolution

By 1942, little more than 70 years had passed since the work of Louis Pasteur in France transformed scientific thinking and revealed to scientists and doctors alike that microbes were at the root of much disease and decay. Inspired by Pasteur, many scientists, including Fleming, had searched for a substance that could kill dangerous microorganisms inside a person's body without harming the body itself.

And some important discoveries had been made. Led by Pasteur's work in the 1870s, scientists had found that microbes could be injected as a vaccine into humans to stimulate their bodies' natural defenses. These defenses would then be armed against active microbes of the same kind, and the patient would be protected from, or immune to, those microbes.

Another important development came in the 1930s, when researchers developed several chemical drugs that could be injected into humans. But these drugs did not always work, and the ones that did were effective only on few specific microorganisms.

The vast majority of microbes could rampage unhindered through the body. Though doctors could name the most dangerous microbes, and even, with the power of their microscopes, see them in people's body fluids, tissues, and blood, very little could be done to combat them.

Killer Infections

In 1942, a large number of the patients who filled the hospitals were children and young people who were suffering from infections caused by dangerous microbes. Frequently these infections led to death.

Women died in childbirth and infants died shortly after birth. Children died from scarlet fever and from infections of the bones, throat, stomach, or brain. The microbes invaded wounds and caused a spreading infection that could kill in a few days. Once infected, a wound as small as a pinprick or a tiny cut could be lethal.

A Thing of the Past

Many of the diseases that kept hospitals filled only fifty years ago are today easily treated with antibiotics—drugs that are made with one microbe that can kill another. Antibiotics were developed thanks to Alexander Fleming's penicillin—the first, and still the best, antibiotic. The contributions of all the people who made the antibiotic age possible were recognized in 1945, when Alexander Fleming, for his discovery of penicillin, shared the Nobel Prize for Medicine with Howard Florey and Ernst Chain—

This illustration depicts a mother weakened by an uncontrollable infection after childbirth.

Top: Fleming was born and spent the first years of his life on this farm in Scotland.

Right: One of Fleming's farming chores was to tend the family's sheep.

the two scientists in Oxford whose work enabled them to give penicillin to the world.

A Childhood in the Ayrshire Hills

Alexander, or Alec, as his family called him, grew up in a place very different from the crowds, noise, smoke, and dirt of London, where he lived and worked for more than fifty years and where he made his lifesaving discovery. He spent the first fourteen years of his life among the windswept hills and glens of the uplands of Ayrshire, in southwest Scotland.

He was born on August 6, 1881, in a remote farmhouse in the hills four miles from the small town of Darvel. Above the farmhouse stretched rough pastureland, and beyond it, the wide, wild expanses of heather-covered moors.

The Fleming family was large. Alec's father had four children with his first wife. After she died, he married again at the age of sixty. Alec was the third of four children his father had with his second wife, Grace. Alec's father died when Alec was seven. His eldest brother, Hugh, helped his mother bring up the family and run the farm.

Alec's early life was comfortable and carefree. The older children were expected to help take care of the animals and do household chores such as fetching water from the spring and fuel for the fire. But the younger boys had little work to do except to mind the sheep. Alec spent most of days in the company of his brother John, two years older than himself, and his brother Robert, two years younger. The three played in the barns, roamed the farm and moors, explored the streams that formed waterfalls and deep cold pools in the glens, and fished in the river waters.

Schooldays on the Moors

When Alec was five, he began to attend a little school on the moors a mile from Lochfield. There, a young teacher taught children from the nearby farms. The students gathered in the single classroom or, in pleasant weather, by the river.

In later life, when he was world-famous, Fleming said that he received his best education in that tiny moorland school, and on the daily walk to and from it. Young Alec learned to be fascinated by nature and developed a keen eye for observing and remembering everything around him. He relied on these qualities throughout his life.

In 1891, when Alec was ten, he and Robert moved to the school in nearby Darvel, a small lace-making town that extended along the floor of a green valley on the Upper Irvine River. The school was a four-mile walk from home, and the boys made the eight-mile round trip daily, in all kinds of weather. At twelve, Alec left behind the small, familiar communities of Lochfield and Darvel to attend a large secondary school in Kilmarnock, a busy industrial town of some 30,000 people.

As John, Alec, and Robert Fleming grew up, their mother and elder brothers, Hugh and Tom, worried about their future. When the boys' father died, he had left the tenancy of the farm to Hugh, who ran it with their mother. Tom had gone to study medicine at Glasgow University. Now the family wrestled with a question: should the younger boys remain on the farm, like Hugh, or should they, like Tom, continue their education and look for other ways of earning a living?

In the end, Tom himself provided the answer. After he qualified as a doctor, he had set up medical practice in 1893 in London, where he treated eye problems. His eldest sister, Mary, had gone to keep house for him. Tom invited John to join them, and found him an apprenticeship in the firm of lens makers that supplied Tom in his work.

In the summer of 1895, Alec joined his brothers when Tom suggested he come to London to finish his schooling. At the age of thirteen, Alec Fleming left the open spaces of Scotland for the bustle, noise, and industrial smoke of London. This enormous city became his home for the rest of his life.

A New Life in London

Six months later, brother Robert joined Tom and Alec. The trio of boys who had together roamed the moorland wilderness were reunited. John learned the trade of lens making, and Robert joined Alec at the Regent Street Polytechnic, which provided classes for a small fee for anyone who wanted to learn.

By the time he was sixteen, Alec had passed all his exams. He had no strong feelings, however, about what kind of work he wanted to do. He took a job in a shipping office of the America Line, a company that ran some of the largest and fastest of the liners across the North Atlantic. As a junior clerk, Alec had to copy documents by hand, keep accounts, and record details of cargo and passengers. He worked carefully and well, but he found the job dull. Young and still seeking a life direction, Alec had not yet found his path.

Inheritance and Opportunity

In 1901, an uncle died and left each of the Fleming children an inheritance. Dr. Tom Fleming at once

Fleming moved to London in 1895 at the age of thirteen.

15

The Fleming brothers found the crowded London streets fascinating after their childhood in the country.

used his money to open a consulting room in London's Harley Street, and he rapidly began to acquire many more patients. By this time, Robert had joined John in the lens making firm. Both of them liked the work, and seemed well settled. When they received their money, they set up their own firm of lens makers, which, in time, developed into a large company with many branches. (Years later, in 1942, a director of this firm, Harry Lambert, helped Alexander Fleming make history.)

Despite Alec's hard work at his clerical job, Tom could see he was not enjoying it. Medicine was proving to be a secure and interesting profession for Tom himself. He wondered if the field would offer better opportunities for his young, unsatisfied brother.

He put the proposal to Alec. Why not use the inheritance to study medicine and become a doctor?

New Directions, New Vistas

Alec liked the idea. It offered a way out of the boredom of the shipping office. Yet he faced a formidable task. He was nearly twenty, several years older than most first-year medical students. He had none of the qualifications needed for acceptance to a London medical school because he had left school at the age of thirteen.

Undaunted, he set out to obtain the qualifications he needed. He found a teacher to tutor him in the evenings, and by July 1901, he took the entrance exam. He passed in all sixteen subjects, and in October 1901, Alec Fleming joined St. Mary's Hospital Medical School.

Fleming adjusted so quickly to his new life that his years as a shipping clerk faded rapidly from his mind. Now he had to investigate the structure of the human body, its tissues and organs, and in the minutest detail. He also had to study how different body parts functioned, although at the time, many discoveries about biological processes had not yet been made.

Fleming had taken up medicine just to escape clerical work. But he soon found he had stumbled into a profession that thoroughly suited him. There was no subject that did not interest him, and he became well known among the students for winning prizes for his work.

Life was not all study, however. He entered the community life of the hospital with great enthusiasm. He played on the hospital's water polo team, joined its dramatic and debating societies, and became a star member of its rifle club. In July 1904,

Fleming (at front of line) and his brothers joined the London Scottish Rifle Volunteers, a part-time military regiment.

when he was twenty-two, Fleming passed his first medical exams. He had decided to become a surgeon.

With the first exams behind them, the students no longer spent their time in lecture rooms and classes. Instead, they frequented the wards and casualty rooms of the working hospital. They trailed through the wards behind a doctor to watch how he examined patients. They had to learn what to look for, and how to piece together a picture of a patient's illness so they could recommend treatment.

Seeing the Limits of Medicine

Although there was a great deal for the students to learn, there was actually very little that doctors in those days could do to treat many of the diseases they encountered in a hospital like St. Mary's.

It had been only nine years since the discovery of X rays, and would be several decades more before

Fleming (front row, second student from left) studied medicine at St. Mary's Hospital.

St. Mary's had an X-ray department. Doctors knew the parts of the body in detail. But exactly how these parts worked together, and what they actually did, was little understood.

Scientists had only just begun to unravel the mechanisms by which the body fights off disease. And no one could really do anything against the vast army of microbes that attacked the human body. The battle against the microbes had scarcely begun.

Of all the places he could have chosen to study medicine, Fleming had unknowingly selected one where some of the most exciting work in this field was being done. One of Fleming's teachers was a pioneer in this new war against disease-causing microbes. His name was Almroth Wright.

The Pioneering Dr. Wright

As a young man in the 1880s, Almroth Wright had studied in Europe under some of the great medical scientists of the time. Microbes had been observed and recorded since the seventeenth century. But the work of Louis Pasteur, beginning in the 1850s, had driven European scientists forward in a great surge of activity. In the 1870s, Robert Koch, a German researcher, first proved that a specific disease is caused by a specific microbe.

In laboratories all over Europe, scientists looked for new ways of handling and studying microbes, particularly the class of microbes known as bacteria. Intricate techniques of staining bacteria were developed, so that the microbes could be more easily watched and tracked.

Wright observed and learned. He took his knowledge back to Britain at a time when few similar developments were altering medical practice there. Later he worked at an army medical school. He became particularly interested in wound infection, and in diseases like cholera, typhoid, and

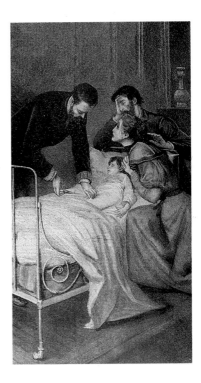

A doctor treats a child for diphtheria, a disease that killed thousands of children annually in Europe.

Almroth Wright was a pioneer in the war against deadly microbes.

dysentery. These illnesses often attacked armies at war: crowding and dirt combined to create a perfect environment for the bacteria that cause these killer diseases to flourish.

No Cures

There were still no cures for many diseases, although Pasteur's work in the 1870s had begun to show how cures might be found. Pasteur had demonstrated that by injecting a vaccine made of bacteria that had been weakened in some way, doctors could force the body's natural defenses to arm themselves. But this idea had not yet been widely accepted in Britain.

Almroth Wright, however, became one of the great pioneers of vaccine therapy, and a fierce campaigner for it. By the time he became professor of pathology and bacteriology at St. Mary's Hospital in 1902, he had come to believe passionately that the right vaccination could do much more than prevent bacterial diseases. He was certain it could also cure them.

Alexander Fleming was one of the students who became caught up in the excitement of Wright's lectures. Wright was an astonishing man, dazzling in style, forceful in ideas, and already possessed of a gigantic reputation for fighting for his cause.

Fleming was greatly impressed by these new possibilities in the war against bacteria. But he did not yet think of becoming a bacteriologist. Surgery was still his main aim, so he took the preliminary surgical exam in January 1905.

A year later, in July 1906, shortly before his twenty-fifth birthday, Fleming passed his final medical examinations. He was now a qualified doctor. He could begin to practice either in St. Mary's or another hospital, or establish a general practice of his own, like his brother Tom.

What Fleming wanted to do was stay in medical school and work for a higher exam to give him more options for the future. He had spent his inheritance, however, and did not have much money. He needed

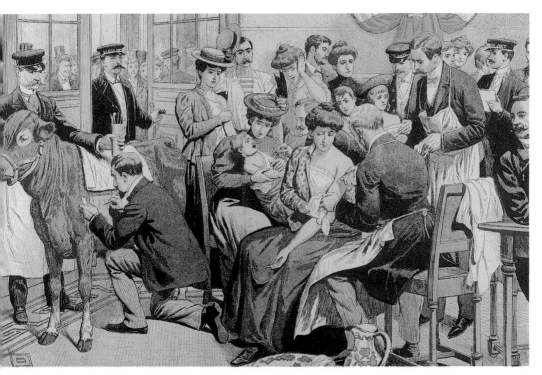

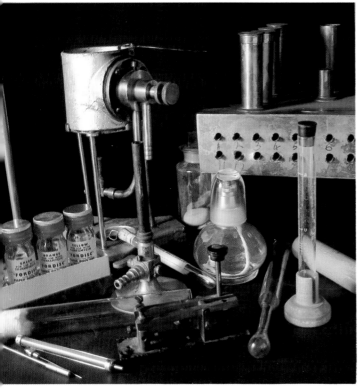

Top: Patients are vaccinated against smallpox. An injection of the microbes that cause cowpox, a relatively mild disease, triggers the body to build immunity against the more serious illness of smallpox.

Left: This photograph depicts the kind of laboratory equipment that Wright and his team used at St. Mary's.

to earn a living in some way and also find the time to continue his studies.

Chance, and the Rifle Club

By chance, there was someone in the medical school who particularly wanted to keep him there—John Freeman, an enthusiastic member of St. Mary's Rifle Club. That year the rifle team hoped to win an important national contest. Fleming was a valuable member of the team, so Freeman set out to keep him busy at St. Mary's, to make sure he would still be around for the contest. Freeman was also a member of Almroth Wright's department, and knew that Wright had a vacancy for a junior assistant. He suggested Fleming for the job.

Thus, in the summer of 1906, Fleming joined Wright's department. The job was meant to be a temporary solution to Fleming's problem. He was soon so engrossed in his work, however, that he remained a member of that department for the rest of his working life.

Fleming worked in this building at St. Mary's Hospital until the 1930s.

Early Work in Bacteriology

Almroth Wright was then 45 years old and at the height of his career. He had an enormous reputation for his pioneering work, much influence, and some powerful friends, including a few in the British government.

His enthusiasm inspired his whole department. Eight or nine young graduates, Fleming among them, took up their work with great energy, convinced they were going to bring about a revolution in medical science.

Wright's team often worked long into the night if crucial measurements had to be taken on an experiment. At teatime each day, they gathered for an animated discussion, and met again around midnight, when they reviewed the day's work. The department hummed with activity.

Fleming fitted in easily, though, at first, quietly. His general good nature and interest in all aspects of the team's research meant he was soon well liked. His practical skills became an enormously valuable resource for the department. Everyone who knew him at this time recalled his deftness and skill with his fingers.

These skills came in especially handy when the team's work involved intricate practical problems, such as how to manipulate or measure minute amounts of substances. Fleming would listen to Wright explain some new idea. Then, without saying a word, he would disappear to his laboratory bench and return after a few hours with some new tiny gadget or an intricate method that exactly solved the problem.

The Body's Natural Defenses

The department was absorbed in investigating what vaccination did to the body's defense system, known as the immune system. What exactly happened when a vaccine was injected? How did the body react? What processes took place that made the body then able to fight off a disease?

Wright believed passionately in the idea that the body's natural defenses could fight effectively against bacteria. To cure bacterial disease as well as prevent it, Wright believed, the body had to be stimulated to make itself invulnerable.

To learn more, Wright's researchers tried to force the immune system to work, and then attempted to observe and measure what was happening. They would, for example, identify the bacteria they believed were causing a disease. Then they would grow the bacteria, kill them, mix them in a fluid to make an injection, and then make sure that each measured amount of this vaccine had exactly the same number of bacteria in it. This process created a standard dose.

The researchers injected vaccines into animals, patients, even themselves. Then they looked at drops of blood from the injected organism under a microscope. They tried to record and measure what had

Fleming studied in the library of St. Mary's Hospital.

23

happened to the blood—how it was different from the blood of someone who had not been injected.

The Swallower Cells

The researchers became most interested in one part of the blood—the white blood cells called phagocytes. Phagocytes are known as swallower cells because they actually swallow and digest bacteria. The team looked at phagocytes in ordinary blood. Then they looked at phagocytes in the blood of someone who had recovered from an infection or had been given one of the vaccines.

For a long time, Wright was convinced that there was a special sort of substance in the blood of these people that encouraged the phagocytes to swallow bacteria. He called this substance opsonin, and the department spent much energy and many midnight hours trying to record and measure what they believed was opsonin at work. During this time, the department was convinced that they were on the verge of a breakthrough in medicine.

The Year of Decision

By 1908, Fleming had passed his next exams with distinction. He decided to work for the specialist examination that would qualify him as a surgeon, and get practical experience in the hospital. By June 1909, at the age of twenty-seven, he had passed the exam. Finally, he was qualified to begin surgical work.

Instead, however, Fleming decided to remain with Wright. Fleming had become totally obsessed about the work. His brother Robert later recalled how the family noticed this shift in Fleming's interest. He stopped talking about surgery and spoke a lot more about Wright's ideas. He even started trying them out on the family. Sore throats, colds, or any other kind of minor infection would prompt Fleming to take a sample of mucus or blood back to the laboratory to look for bacteria. He usually reappeared with a vaccine that he would duly inject into the patient.

Fleming used this microscope in his experiments.

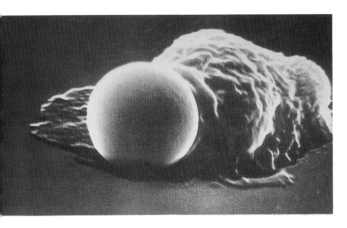

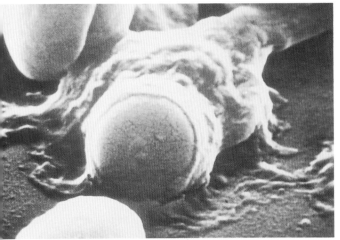

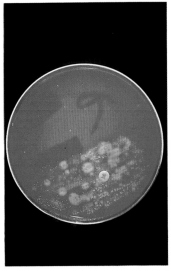

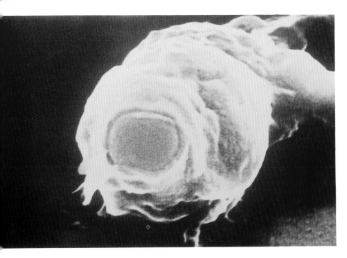

Top: In this culture dish, the pale areas are red blood cells that have been dissolved by bacteria.

Left: A white blood cell— a phagocyte—approaches, surrounds, and consumes a red blood cell. Phagocytes also remove foreign particles such as microbes from the body.

Fleming abandoned his plan to become a surgeon and decided instead to continue his research work with Almroth Wright.

Dr. Fleming's Reputation Grows

As his work continued, Fleming gained a reputation as an expert on vaccine therapy. He also made a name for himself in one particular field. He developed a new test for the devastating disease of syphilis, a test that used just one or two drops of blood.

Untreated, syphilis destroys the body slowly over many years, and can be passed on by a mother to her unborn child. In Fleming's day, it was particularly difficult for doctors to tell if someone was suffering from syphilis, because the spiral-shaped bacteria that cause it were difficult to isolate. Fleming's test allowed doctors to detect the presence of the bacteria based on changes the bacteria caused in the blood. The improved test was a major advancement, and many patients came to Fleming for it.

A Wonder Drug

In addition to creating the new syphilis test, Fleming became one of the first to use a new and better treatment for the disease itself. In Germany, a scientist named Paul Ehrlich was trying to make chemical substances that would be more poisonous for microbes than for people. He wanted to find a "magic bullet"—a chemical weapon that would attack only the microbe it was intended to fight.

Ehrlich tested 605 chemicals and found one that was extremely poisonous to the bacteria that cause syphilis, but much less harmful for people. He called it 606, and later, salvarsan. It could be given by injection so that it reached every part of the body and found the bacteria wherever they lurked.

Almroth Wright knew Ehrlich, and was one of the first to get a supply of salvarsan for testing in Britain. He asked Fleming and another doctor in the department, Leonard Colebrook, to try it out. They used it to treat syphilis with great success, and Fleming became known an authority on treating the problem.

World War I

Fleming's work with syphilis was abruptly interrupted in 1914. World War I had begun, and Fleming was transported many miles from the familiar walls of St. Mary's to a hospital in France.

Almroth Wright had fought tirelessly for his typhoid vaccine to be given routinely to soldiers. With the outbreak of this war, he campaigned even harder. He insisted that all troops be injected against typhoid immediately. In addition, he demanded that wound infections be treated with the new vaccines he was developing. Wright still believed that weakened bacteria should be injected into wounded patients to stimulate their own immune system, which would fight infection.

The army accepted Wright's views about the typhoid vaccine, and gave it to the soldiers. Without the vaccine, there would probably have been at least 120,000 deaths in the British Army from typhoid over the four years of World War I. Instead, there were only 1,200.

Army doctors were not convinced, however, that Wright was correct about vaccines for wounds. They suggested he set up a research unit at a war hospital to investigate further.

The Casino in Boulogne

In October 1914, Wright, Fleming, John Freeman, Leonard Colebrook, and several other bacteriologists from St. Mary's found themselves surveying the grim misery of wounded men from the battlefields. The hospital in which they worked was located in the grand casino in Boulogne, France. Beneath the casino's elegant chandeliers, rows of camp beds held groaning, bleeding soldiers. Many waited days for surgery. Their wounds were sickening, filthy, and often infected. The culprits were a variety of common bacteria, but sometimes the dreaded gas gangrene had taken over.

This cartoon depicts Fleming wearing the uniform of the London Scottish Rifle Volunteers and holding a syringe of 606. Fleming became well known for using 606 to treat syphilis.

27

Conditions in the trenches of World War I—dirty, bloody, overcrowded, and damp—allowed deadly microbes to multiply and spread rapidly.

Gas gangrene was caused by a virulent microbe that spread with terrifying speed. It poisoned the wound, blackened the surrounding muscles and skin, and filled the area with gas in a few hours. The only treatment was to cut off the limb immediately. Delay meant certain death.

The Need for Antiseptics

In cases where gas gangrene had not yet occurred, the doctors could only pour chemicals known as antiseptics into the wounds to try to kill the bacteria. They had been treating wounds in this way since the Scottish doctor Joseph Lister introduced the first antiseptic in the 1860s.

When Lister read about Pasteur's work in 1867,

28

Soldiers wait for treatment at a first-aid station during World War I.

he decided that bacteria must be causing the infections that usually followed surgery in hospitals. He began using a substance called carbolic acid to keep equipment, air, and surgical wounds free of bacteria. He transformed surgery, not just because other doctors began to use carbolic as an antiseptic, but also because they began to understand the need for cleanliness.

But when it came to killing bacteria in an already-infected wound, the problem was not so easy to solve. How could you apply antiseptics and not damage the tissues of the body? Lister tried diluting carbolic acid so as not to do a great deal of harm to anything but the bacteria, but for years he went on searching for a chemical that would destroy bacteria without harming living human tissue at all. He never found it. Over the next seventy years, many other workers launched on the same quest. But none had been successful.

As Wright's unit began work in Boulogne, general practitioners and army doctors alike still had an unshaken faith in chemical antiseptics for treating wounds. Despite their conviction, the liberal application of these chemicals did not usually achieve the desired result. The infection invariably continued, or even got much worse. And no one knew what these substances were doing to the body tissues, particularly deep in the wound.

Important Evidence

Why didn't antiseptics work? Which bacteria were causing infections? What was happening in the blood during infection, and what then happened when antiseptics were poured on? And most importantly, why did antiseptics that worked in a test tube in the laboratory not work in a real wound?

Wright, Fleming, and the others set out to answer these questions. They gathered information from everywhere as they tried to get a clear picture of the infection process. They found out, for example, that the men's clothing was the worst source of bacteria.

And they made a particularly important discovery. In fresh wounds, or untreated ones, the seeping fluids and blood were full of phagocytes, all busily consuming bacteria. But in wounds where doctors had used antiseptics, there were very few phagocytes. Those that the doctors could see were dead or dying. But the bacteria, which the antiseptics were supposed to kill, were still very much alive and multiplying rapidly.

Here, before their eyes, was evidence at last of something they had all believed for years. The body's own phagocytes were vital to fighting off bacteria. And the antiseptics killed the phagocytes.

Fleming's Artificial Wound

Fleming had another idea about why antiseptics were ineffective at treating wounds. He suspected that antiseptics on the surface couldn't work because they literally couldn't reach into all the torn channels of the injuries. To test this theory, he made an artificial wound from a test tube. He heated the tube to soften the glass, and then shaped it into hollow spikes, like the tunnels of a jagged wound. He filled this with infected liquid, emptied it, and refilled it with antiseptic. His experiment showed that the bacteria started spreading again because the antiseptic had never reached deep into the spikes to kill them all.

With all the evidence his team had collected, Wright began his campaign to change the treatment of wound infections. Antiseptics should not be used, he insisted. The wound should only be rinsed with a strong salt solution, which would encourage the phagocytes to do the main work. The wound should then be protected from new bacteria by clean dressings. These ideas, however, were years ahead of current medical thinking. Therefore, the work of the dedicated team of bacteriologists in Boulogne did not change most doctors' treatment methods in World War I.

The Killer Influenza

The years of war work made an indelible impression on Fleming and all of his fellow researchers. They struggled to conquer the infections, but one impossible fact remained. They could do nothing to save most of the wounded soldiers they treated.

Just before the end of the war, an epidemic disease made a dark and painful time even worse. In 1918, an influenza epidemic swept through of Europe. Doctors were helpless to stop it, and more than twenty million people died. Young, healthy people felt slightly ill one day, and were dead the next. Far

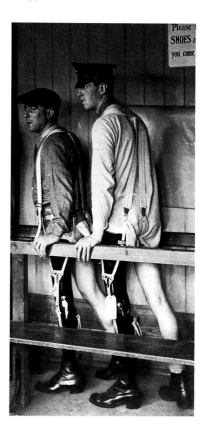

Amputation was often the only way to save a soldier who was infected with gas gangrene. These soldiers' amputated legs were replaced with artificial limbs.

31

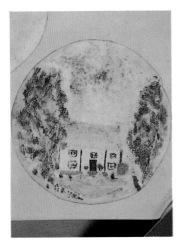

Fleming created these illustrations with bacteria that change colors as they grow.

more people died of influenza than were killed in the war itself.

Military hospitals were filled with victims. Wright's unit, particularly Fleming, struggled to work out why a usually mild disease had suddenly become so deadly. There were no obvious answers yet, and again the same devastating fact remained: doctors had no way of killing bacteria without also killing the body's tissues and phagocytes.

Sareen

As the influenza epidemic subsided and the war came to an end, so did the team's time in Boulogne. In January 1919, Fleming finally returned to the inoculation department at St. Mary's Hospital. But it was not a return to his old way of life.

During one of his brief periods of leave from Boulogne, Fleming had done something so completely unexpected that at first no one believed it. He had to show acquaintances a photograph to prove it was true. He had gotten married.

Before the war, he had become friendly with two sisters, twins named Sally and Elizabeth McElroy. They were nurses who ran a private nursing home. Sally, who later became known as Sareen, was an energetic, outgoing person who talked and laughed a lot. She was so different from Fleming—who was well known for never saying much and listening a great deal—that friends decided the couple's differences were probably what drew them together.

The two were married in London on December 23, 1915, although Fleming, now thirty-four, returned to France immediately after the wedding, and Sally continued to run the nursing home. Not long after, Sally's sister Elizabeth married Fleming's brother John.

Fleming's married life truly began in January 1919, when he returned to England. In 1921, he and Sareen bought the Dhoon, a country home set among fine trees, and—much to Fleming's liking—a large garden with a river where they could swim, boat, and fish. From then on, the Flemings spent many weekends improving their large home. When their son Robert was born in 1924, Sareen began to spend more time there, often joined by the children of Fleming's brothers Robert and Tom. The Dhoon became a major focus of the couple's life together, and they spent many happy years there.

The First Great Discovery

Fleming's marriage, did not distract him from his search for the perfect antiseptic. He had been busy growing a wide variety of microorganisms and observing how they behaved—paying particular attention to how they reacted to different substances. In 1921, Fleming made his first great contribution to medicine—the discovery of the natural antiseptic lysozyme.

Fleming and his wife, Sareen, bought this country house, the Dhoon, in 1921.

There was a standing joke between Fleming and a young doctor who was working with him. Dr. Allison, Fleming teased, was much too neat and tidy. He cleared up his bench at the end of each day, busily discarding growths of bacteria he no longer wanted and cleaning the dishes for further use.

By contrast, Fleming's bench was crowded with bacteria growths (known as cultures) from experiments that he had performed over many weeks. He liked to leave them for a while and then have a good, long look at everything before he dumped them into disinfectant. You never knew, he thought, when something interesting might happen.

One day, he was looking through the piles of old culture dishes, preparing to clear them away. Suddenly he stopped. He peered at one carefully for some minutes, and then showed it to Dr. Allison, saying no more than, "This is interesting."

Fleming had been suffering from a cold some weeks earlier, and in his never-ending search for greater understanding of bacteria and disease, had started to grow a culture from a blob of the thick

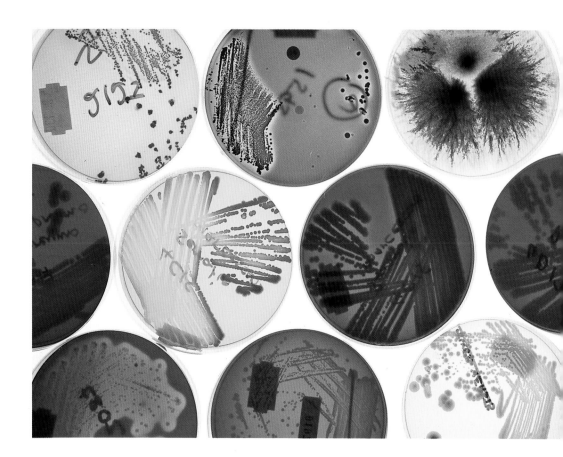

Researchers grow colonies of different bacteria in culture dishes. Bacteriologists must identify the ones that can cause infection.

fluid from his nose. It was this dish that he showed to Allison.

But what had caught his attention, and what Allison now saw, was that there were also colonies of golden-yellow bacteria growing on the dish, everywhere except immediately around the blob of nose fluid. Next to this they had started growing, but had become glassy and seemed to be dissolving. They were growing normally only some distance from the nose fluid.

Despite his few words, Fleming was greatly excited. He had seen the significance. Could there be something in the nasal fluid that actually killed bacteria?

Mysterious Component

Quickly and carefully, Fleming tested fresh nasal fluid. The same thing happened: the bacteria would not grow near it. Fleming wondered if this happened

Researchers use various methods to grow and test bacteria. In the picture at left, a colony of bacteria is visible on the end of the loop.

with other people's nose fluid. He pestered friends and colleagues for samples. The same thing happened with their cultures—and it happened even when they didn't have a cold.

Fleming moved on to other body fluids. He insisted that people cry for him. When they did, he hurried away with their tears; he tested saliva, pus, and blood serum (the clear fluid that seeps from a blood clot). All of these substances had this amazing ability to stop these golden-yellow bacteria from growing!

For years, Fleming had followed the trail of research blazed by Almroth Wright. But here was something new, something unknown, that he had found himself. Was this substance some previously unrecognized part of the body's immune system, something that worked together with the phagocytes to combat bacteria?

During the next weeks, Fleming filled the pages of his notebooks with more experiments. He wanted to know which body fluids and which parts of the body contained this bacteria-dissolving substance. He wondered which bacteria it would kill.

He found that the substance seemed to be everywhere. Skin, mucus membranes, most internal organs and tissues, and even hair and nails had it. He also found it in some animals, and in plants, flowers, and vegetables.

He also found, however, that it affected only a few bacteria, and most of the dangerous ones were barely affected at all. Today, scientists know that Fleming had discovered the body's first line of internal defense. If the bacteria could get past it, then the phagocytes had to move into action.

The substance was, in fact, the body's natural antiseptic. At the time, however, its full significance was not understood, and it took many more years of research before the processes of this defense system were properly recognized. And because the substance did not deal with the most dangerous bacteria, scientists were little interested in it at the time.

Lysozyme was the name Fleming chose for the new substance, and he continued investigating it for many years. Along with Allison and other young bacteriologists who worked with him, he published many reports about lysozyme. But no one really grasped the significance of it. Unlike Wright, Fleming was not a man to force anyone's attention.

Rather Interesting

Those who worked with Fleming during this time remember him usually bent busily and happily over his laboratory bench. His door was always open to visitors and gossipers. Often he would stroll into the main laboratory to see what was going on among the younger scientists, and chat about the work or about some general topic of scientific interest.

Then, on a September morning in 1928, Fleming wandered, as usual, into the main laboratory. He was

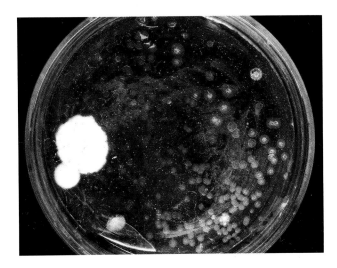

This culture dish, the original one on which Fleming first observed penicillin, is housed in the British Museum. The white area is the colony of penicillium. To its right, the colonies of staphylococci bacteria are dissolving and thus appear transparent. At the far right, staphylococci colonies out of the reach of the penicillin grow normally.

holding a culture dish, and had an air of carrying something rather interesting. Most of the other scientists assumed that he was simply showing them another example of lysozyme—this one from a mold—at work. Not until years later would the young scientists in the St. Mary's laboratory realize that on that day, they got their first look at penicillin.

Miracle Mold

Fleming had seen something that escaped the others who peered at his culture dish: the mold had done something that the familiar lysozyme had never done. It had attacked one of the most common and most dangerous bacteria—the round, virulent cocci, known as staphyhlococci, that grow in clumps like bunches of grapes.

Over the years, the story of Fleming's discovery has been told, retold, and often embellished. The actual events of that day, however, were not so different from Fleming's discovery of lysozyme. Once again, he had been sifting through old culture dishes before cleaning them. On that historic occasion in 1928, he was chatting with an assistant named Pryce.

Poisonous!

Pryce had left the department to take up other research. But he dropped in to see how Fleming's work was progressing. The two had been growing staphylococci from boils, abscesses, and nose, throat, and skin infections. Then they had left the bacteria at room temperature to observe how the microorganisms changed and whether the changes altered their strength.

Fleming was clearing away the cultures of several weeks back. He had inspected and piled most of the dishes into disinfectant for cleaning. He picked up one, not yet covered by disinfectant, from the top of the pile to show to Pryce. As he looked at it again, he paused. He murmured simply, "That's funny," and passed the dish to Pryce.

Pryce saw the usual smooth, dome-like colonies of golden-yellow staphylococci. They covered the dish—except at one side. There, near the edge, a patch of fluffy mold had started to grow, and near it, the colonies of staphylococci were transparent. Close to the mold, there were no colonies at all.

Like the other people Fleming showed the moldy plate to that day, Pryce thought little about what he had seen. Fleming, however, was already very excited. Something had killed staphylococci! He photographed the dish and preserved it. (The historic dish is today housed in the British Museum.)

Fleming lost no time in trying to duplicate what had happened. He put some of the mold onto a

Fleming grew cultures in this dish to show that bacteria can protect themselves from penicillin by producing a substance called penicillinase. On the top half of the dish, the staphylococci grew where Fleming placed both penicillin and penicillinase. At the bottom, where there was no penicillinase, the penicillin stopped the staphylococci from growing.

fresh dish, grew it, and then tried to grow staphy-lococci alongside it.

He couldn't. There was no doubt—something seeping from the mold was poisonous to staphylococci.

Harmless!

Over the years, Fleming had tested many substances that could kill bacteria. Now he followed this well-worn path even more eagerly. One by one he did the same battery of tests on the mold juice. Did it kill other bacteria? Did it harm the phagocytes of the blood? Did it damage the delicate tissues of the human body?

The results were astounding. The mold juice could stop some of the most dangerous bacteria from growing. An unknown mold, landing by chance on a culture dish, could apparently simply dissolve them.

And the mold did not harm the body—the phagocytes were still busily at work. Even when the juice was injected into the body of a mouse and a rabbit, it produced no ill effects. Even when the juice was heavily diluted, it was lethal for the virulent bacteria. Fleming had found something that was vastly more powerful against bacteria than carbolic acid, yet harmless to the delicate living blood cells.

Later, however, he discovered some serious problems. Most chemical antiseptics killed microbes within a few minutes. The mold juice took several hours. And it seemed to lose its power completely in mixtures that had blood serum in them. For Fleming, these findings were very disappointing. They meant that, in wounds or infected areas that seeped serum, the mold juice would lose its power long before it could kill the bacteria.

Penicillin

Fleming consulted another scientist who knew about molds. His colleague told him that Fleming's mold was one of the penicillium group. In February 1929,

The powerful penicillium mold that Fleming observed is known as penicillium notatum.

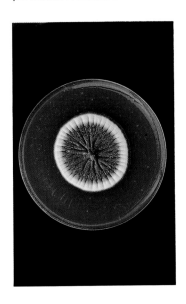

Fleming and his assistants discovered that the penicillium notatum *mold grew best on the surface of a kind of meat soup at room temperature.*

Fleming began using the name *penicillin* for his bacteria-killing substance.

Fleming became a fanatical collector of molds. He could not, at first, believe that his was the only one with this extraordinary power. His mold quest became a source of bemusement among family and friends. Cheese; jam; worn clothes, boots, and shoes; old books and paintings; dust and dirt of all kinds, from Fleming's own home and from friends' homes—nothing was safe from the scientist's hunt for interesting scrapings he could carry back to test in his laboratory.

Fleming's mold, however, was the only one he found with amazing antibacterial properties. The more Fleming found out about it, the more extraordinary it seemed. It could even kill the bacteria that caused the hideous gas gangrene he fought so desperately and so ineffectively against during World War I.

Trying to Purify It

Fleming asked two assistants, Ridley and Craddock, to produce mold juice for his experiments. They grew the penicillium in a kind of meat soup in large bottles with flat sides. Over several days it spread over the surface of the soup in a fluffy layer, while the liquid below turned yellower and yellower, and became more and more powerful in its bacteria-killing effect.

But the scientists had difficulty trying to isolate the penicillin from the liquid and other things in the juice. And when they did partly succeed, there were new problems: the penicillin very easily lost its bacteria-killing powers.

In the early months of 1929, however, one thing in particular caught Fleming's attention. Penicillin killed many bacteria, but not all. It did not harm the bacterium that scientists believed at the time was the cause of influenza. This microorganism was called Pfeiffer's bacillus, and it was difficult to isolate.

Fleming at once saw an important use for penicillin. He could use it to purify influenza vaccine, like a weed killer, killing all bacteria except the Pfeiffer's bacillus he wanted to isolate and cultivate. The memory of the great influenza epidemic and its 20 million victims was only ten years old. The desire to find for a weapon against influenza was ever-present in Fleming's mind. He also discovered he could use penicillin to isolate other bacteria, such as those that cause the childhood disease of whooping cough.

This use, in the laboratory for selective growing of chosen bacteria, allowed penicillin to emerge as the first great antibiotic drug. Fleming kept his penicillium mold going, and produced juice week after week for vaccines. Scientists around the world asked for samples to use in isolating the difficult influenza bacteria. The mold became established in other laboratories, busily performing its bacteriological work.

Disappearing Powers

During the next few years, there were two other attempts to extract penicillin and find out more about it. Both attempts were stalled by the problem that Ridley and Craddock had already faced: at a certain point in the effort to purify the mold, its bacteria-killing powers seemed simply to disappear.

By 1938, two other scientists, Howard Florey and Ernst Chain, had begun working together in Oxford to investigate Fleming's other discovery—the body fluid antiseptic, lysozyme. When that work ended, they turned to penicillin.

The climate of science was now very different from what it had been only a decade before when Fleming had first discovered his mold. In the years between Fleming's observation in 1928, and the

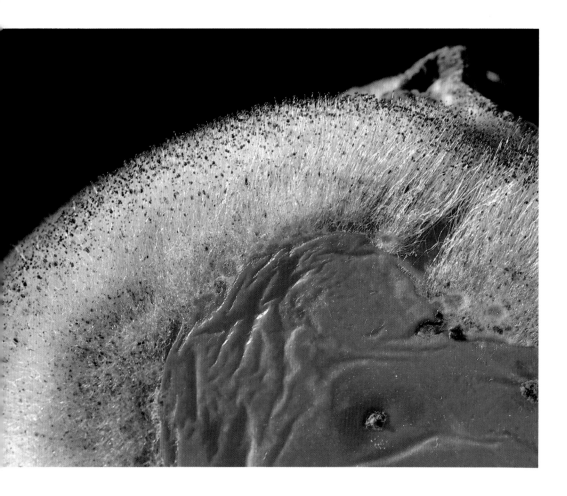

beginnings of Florey and Chain's work on penicillin eleven years later in 1939, scientists in Germany had made a gigantic, revolutionary advance in the treatment of bacterial disease.

The Sulfonamides

Ever since Ehrlich had developed salvarsan in 1910, scientists had been trying to produce chemicals that would kill bacteria by injection into the body, just as salvarsan killed syphilis bacteria. In 1935, a scientist named Gerhard Domagk announced he had discovered a group of dyes that, taken by mouth or injection, protected mice from the killer streptococci. The one that worked best was a rich red dye called Prontosil. Prontosil proved just as effective on humans as it had been on mice, particularly for

Top: Common molds growing on an orange (left) and a tomato (right).

Opposite: Penicillium spores grow in long chains. Each spore can break off and become a new organism.

treating puerperal fever, a severe infection that struck and frequently killed women after childbirth.

The moment that news of Prontosil broke, other scientists began trying to make similar substances, and it was not long before a group of drugs known as sulfonamides was developed. Some of them worked well. People who suffered from dangerous streptococci infections that doctors were previously unable to treat, and that often proved fatal—including scarlet fever, pneumonia, ear infections, and meningitis—could now be saved.

Prontosil and its relatives were still limited. There were problems with the sulfonamides. They did not attack all bacteria, and those bacteria that were attacked could develop a resistance against the drugs. And even the best of the drugs could have side effects that ranged from skin rashes and violent vomiting to death.

Howard Florey

Howard Florey became professor of pathology in the William Dunn School of Pathology at Oxford University in 1935, the year of the sulfonamide revolution. He was an Australian medical scientist who had come to study at Oxford in 1921, at the age of twenty-three. There, he had quickly earned a substantial reputation for the quality and imaginative range of his research on the detailed internal functioning of different parts of the body.

In 1929, he had become interested in finding out exactly how the stomach worked. He was especially intrigued by the power of the stomach fluid known as mucus to kill certain bacteria. He remembered that in 1922, Fleming had reported the discovery of the natural antiseptic lysozyme in body fluids, including stomach mucus.

At once, Florey became intensely interested in finding out precisely how lysozyme worked, and this line of research lasted eight or nine years. In 1938, it brought Florey into a scientific partnership with Ernst Chain, another young scientist.

Howard Florey was the head of the Oxford team that produced penicillin in usable form.

By this time, the sulfonamide revolution had given scientists new ideas. They had begun to understand that infections could be best attacked not by pouring antiseptics directly onto the infected tissues of the body, but by injecting bacteria-killing substances straight into the circulating blood—that is, by using them systemically.

This new understanding flourished among doctors and medical scientists alike. It launched them on an even more energetic search for better systemic weapons in the war against bacteria.

The Florey and Chain Team

In 1938, as their lysozyme work drew to an end, Florey and Chain pursued the study of other natural bacteria-killing substances. Chain searched scientific publications from all over the world and found about two hundred reports on bacteria that scientists had seen stop other bacteria from growing. One was Fleming's 1929 report on penicillin. Florey and Chain chose three bacteria to investigate, including penicillin.

Ernst Chain, a scientist on Howard Florey's team, succeeded in producing a penicillin that was pure enough to be effective when injected into the human body.

They acquired a sample of the penicillium mold from a laboratory along the corridor at the Dunn School. It was being used there, as in many other labs, for isolating selected bacteria. The two scientists set to work to find out more about the mold.

Soon, they faced the same problem that had defeated earlier attempts to separate the bacteria-killing substance from everything else in the mold juice. At a certain point in the process of isolating and extracting it, the penicillin seemed simply to disappear.

Dark Days at Oxford

Despite their frustration, Florey and Chain did not give up. Before long, they had transformed the William Dunn School of Pathology from a teaching and research laboratory into a penicillin production factory. Oil cans, food tins, trash cans and bathtubs, hospital bedpans, milk churns, coolers, and library bookracks—all were appropriated for the two scientists' mammoth effort to produce enough mold juice to fuel their work. In the first dark months of World War II, as the people of Britain dug air-raid shelters; suffered food, fuel, and clothing rationing; and evacuated from cities to areas less likely to be bombed, the Oxford team began their research.

By the middle of March 1940, Chain had won his first penicillin battle. He had extracted enough to start testing it on animals. He had 100 mg in the form of a brown powder that was much stronger than Fleming's crude mold juice. To the scientists' enormous excitement, the more concentrated form of the mold still did not harm animals, phagocytes, or living tissues in the body.

By the end of May, Florey was ready to do one experiment that Fleming had not done. He had to find out if injected penicillin could cure a fatal infection in animals. He knew that penicillin took more than four hours to kill bacteria in a test tube. He knew it was passed out of an animal's body, in the urine, in only about two hours. But for Florey, the

This diagram depicts the process by which the Oxford team used everyday objects to purify penicillin.

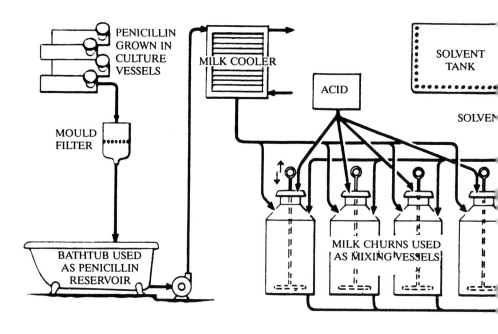

only option was to try: he prepared the crucial test to see if the drug could work systemically, despite this time lag.

The Crucial Experiment

At 11 o'clock on the morning of Saturday, May 25, 1940, Florey injected eight white mice with a deadly dose of streptococci. Four were put back into their cages. Two of the other four were injected with single doses of penicillin, and the remaining two were given five smaller doses over the next ten hours.

By the following morning, the four mice that had not been treated with penicillin were dead. The four that had received penicillin were very much alive.

The scientists repeated the tests. They did different tests. They performed endless experiments to find out what penicillin could and could not do in the body. They investigated how precisely it should be given, how often, for how long, and in what doses. They spent long weeks in which Florey and

These cylinders contain antibiotics developed with the same methods used at Oxford.

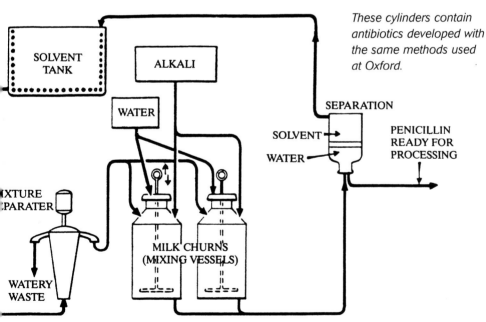

his assistant, James Kent, slept in the laboratory and woke every three hours to inject one batch of animals, inspect others, and record every detail.

By July, the experiments were complete. The scientists reported their results in the medical magazine, the *Lancet*, on August 24, 1940. Few people recognized it at the time, but the article was a signal that a new era in medicine was dawning.

From Mice to People

For the Oxford team, the transition from mice to people was the obvious next step. Immediately, however, they faced an enormous challenge: producing enough penicillin to perform their experiments. Human beings and mice are greatly different in size. To treat a person, the scientists would need 3,000

Citizens took shelter from bombs in the London subway during World War II. Florey and his team continued their research throughout the war.

times the dose of penicillin used to treat a mouse. Unfortunately, vast amounts of mold juice were needed to produce even tiny amounts of penicillin. Florey and Chain calculated that they must produce five hundred liters of mold juice each week for several months to get enough penicillin to treat five or six patients. How could they possibly produce such a quantity with their makeshift equipment?

It seemed an impossible hurdle: until the researchers had enough penicillin to prove its effectiveness on people, no drug firm would mass-produce it. And without help from a drug firm to mass-produce it, the scientists might never have enough to do the critical human tests.

Top: These bottles contain penicillin in brown powder form.

Bottom: The team at Oxford used bathtubs and oil cans in the production of mold juice.

At the Oxford lab, "penicillin girls" sprayed the penicillium mold into the culture vessels under controlled, germ-free conditions.

Still, the scientists pressed on. They turned the Dunn School into a factory: in one classroom six so-called penicillin girls used paint sprayers to distribute mold spores in the culture vessels. In another room, the vessels were incubated at just the right temperature while the juice developed. In still other rooms, the seven scientists and ten assistants of the Oxford team worked day and night to extract and purify enough penicillin to treat a handful of patients for a few days each.

Fleming Goes to Oxford

Fleming learned of the Oxford penicillin work when he read about the animal experiments in the *Lancet*. He lost little time in going to see the operation for himself. On the morning of September 2, he arrived at the Dunn School. With his usual economy of words, Fleming said little and observed a great deal. Florey showed him around, explained everything in great detail, and gave Fleming a sample of their penicillin.

Back in London, Fleming in turn sent Florey some cultures of penicillium mold that produced a good yield of crude penicillin. He wrote enthusiastically, "It only remains for your chemical colleagues to purify the active principle, and then synthesize it, and the sulfonamides will be completely beaten."

By the beginning of 1941, the factory in Oxford had finally produced enough penicillin to plan its first human test. The penicillin was twice as strong as the samples they had used on animals.

The First Human Patient

On February 12, 1941, the team treated their first patient: policeman Albert Alexander, who had been infected by a rosebush scratch. Staphylococci and streptococci had overwhelmed his face, scalp, and eyes. Massive doses of sulfonamides had not helped.

Margaret Jennings was one of the women on the Oxford research team.

Within twenty-four hours of the first penicillin injection, Alexander's condition showed a dramatic, unmistakable improvement. The bacteria in the policeman's body, however, were not completely conquered. After a period of steady improvement in his health, the bacteria began to multiply again. This time, there was no penicillin left to treat him. Albert Alexander died on March 15, tragic proof that penicillin must be maintained in the body long enough to do all its work.

The Second Patient

The second patient was a fifteen-year-old boy dying of infection after a hip operation. With penicillin, he recovered completely. Six more patients treated with penicillin left the researchers in no doubt of the drug's effectiveness. Every patient improved dramatically; two recovered from almost certain death.

But even these successes, Florey knew, were not enough to convince the world: he felt that at least one hundred patients must be treated. Yet the scientists' experiments had shown them that about two thousand liters of mold juice were needed to treat just one severe case of infection.

Florey was still unable to get help from British drug firms, so he sought help in the United States.

There, he managed to stimulate interest enough for production of mold juice to begin in an agricultural research laboratory in Peoria, Illinois.

The Oxford team, however, never got the penicillin from these efforts. In December 1941, the United States entered the war. At once the Americans recognized the importance of penicillin in treating battle wounds, and all production was geared to the American war effort for battle casualties.

Once again, the Oxford scientists were left to their own efforts. They prepared to collect enough penicillin for a second set of medical tests. These tests, performed during 1942, proved penicillin's miraculous powers beyond doubt. Fifteen patients had infections so serious that doctors had given up on them. Treated with penicillin injections, however, all but one recovered completely. In that one case, the bacteria developed a resistance to penicillin, and the patient died.

Fleming Sees the Miracle

In August 1942, Alexander Fleming saw for himself what penicillin could do. He himself injected Harry Lambert with penicillin at St. Mary's Hospital, and saw the man return from the brink of death.

Until the case of Harry Lambert, Fleming had not observed firsthand the evidence of the Oxford penicillin at work. Soon, however, he had additional opportunities to witness its effectiveness. In 1943, St. Mary's received a supply of penicillin from Oxford to try out on patients.

George Bonney, Fleming's resident medical officer, or house surgeon, worked alongside Fleming for the trials. The first case was a little girl. She was suffering from an acute infection of the bone by staphylococci. Her temperature had soared to 106 degrees F (40 degrees C), and she was dying.

Years later, George Bonney remembered the cloudy yellowish fluid of the massive penicillin injection the doctors had to give her, every three hours, throughout day and night. "She was plainly dying," he said. "The next morning she was plainly going to live. It was the first time I'd seen such a thing. I shall never forget it. It was an absolute miracle."

In addition to penicillin's healing properties, the doctors were able to witness the drug's other great miracle. The doctors were all accustomed to seeing the sometimes severe side effects of the sulfonamides. Yet penicillin did no harm at all. It could be pumped into the body, or poured onto open wounds, and there would be no problems. One severe eye infection that occurred in newborn babies had always been untreatable, and could lead to complete blindness. Yet now, doctors could simply drop penicillin onto the eye and watch as the grotesque swelling and seeping pus simply disappeared.

Care and Caution

Even after successes like these, Fleming never lessened his enthusiasm or care as a bacteriologist. He had years of work on antiseptics and sulfonamides behind him; he knew a lot about the infinite capacity of bacteria to adapt and develop resistance to the weapons used against them. Thus, from the beginning, he was aware of the potential for such dangers in the use of penicillin.

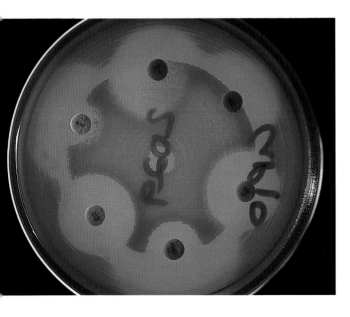

The cultures in this dish show how various antibiotics affect bacteria. The dark spots are paper discs that are each soaked in a different antibiotic. The antibiotics that have clear areas around them are the ones that are most effective against the bacteria.

Penicillin, he was determined, must be carefully and thoughtfully used. It must not help resistant strains of bacteria to develop. Before Fleming allowed penicillin to be used on an infection, doctors had to first confirm that the particular bacteria were vulnerable to the drug. Only then were doctors permitted to use it on a patient.

Even in those early days, Fleming predicted the wider use of penicillin in surgery—particularly wound surgery—to wipe out infection. By 1944, these predictions were already coming true. Doctors were treating war casualties with penicillin. Fleming's memories of the ravages of poisoned wounds in World War I were vivid. Now he saw men who remained free of infection days after their injuries.

Fame

Soon after Harry Lambert's cure in the summer of 1942, news of Fleming's miraculous treatment had

During World War II, doctors treated wounded soldiers with penicillin to ward off infection.

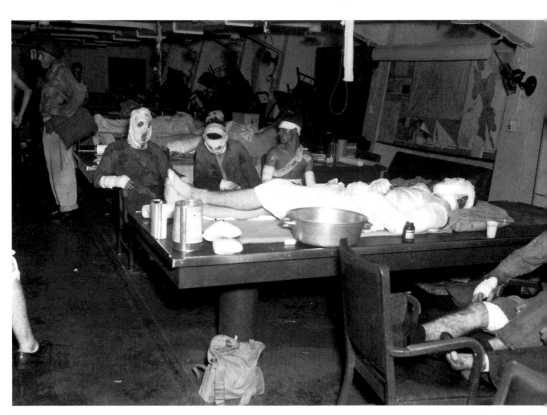

reached the newspapers. The tremendous results in Oxford had heretofore attracted little publicity, but now the story broke. Reporters, hungry for good news in the grim years of the war, besieged St. Mary's in an effort to learn more about this wonderful new substance that could save lives.

From that moment on, Alexander Fleming was no longer a private individual. He shot to fame, and the story of penicillin's discovery, enriched by the tale of Fleming's boyhood in the wilds of Scotland, captured the public imagination. The story was also frequently embellished by fantasy: an article in one paper reported that Fleming had found penicillin when he ate moldy cheese that cured a boil on his neck! Another, ignoring the years between 1928 and World War II, told the tale of bomb dust carrying the miracle into Fleming's laboratory.

Florey disliked this kind of publicity. He also feared that it would create a demand for penicillin

"The great strides in understanding natural phenomena are the result of the labours of thousands of people, some of whom are good scientists—and some not so good. Their combined labours might be likened to the Pointillists who applied little dabs of colour to the canvas and built up a beautiful picture. Scientists can, with luck, from time to time, put a nice dab of colour on a metaphorical canvas; but, for the elaboration of the finished work, they are dependent on the activities of thousands of colleagues."

Howard Florey, 1966

In 1945, Alexander Fleming, Howard Florey, and Ernst Chain were jointly awarded the Nobel Prize for Medicine.

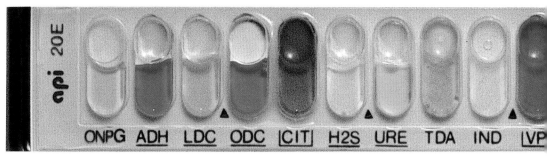

ONPG   ADH   LDC   ODC |CIT| H2S   URE   TDA   IND |VP

Right: In the method shown here of testing for bacteria, different substances are added to samples of blood serum in the plastic wells, and the resulting changes are analyzed.

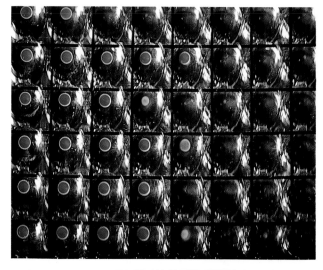

In the test shown here, each well is filled with a different antibiotic. The drugs' effectiveness is revealed by the size of the bacteria-free zone around each well.

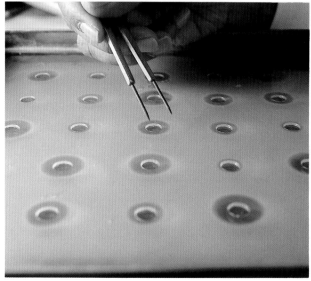

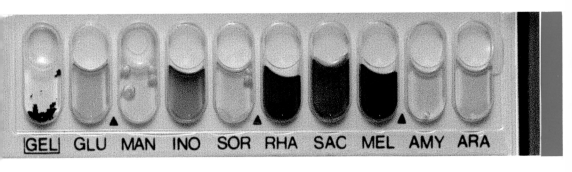

| GEL | GLU | MAN | INO | SOR | RHA | SAC | MEL | AMY | ARA |

among civilians before he could produce enough to supply them. He turned the reporters away, but the name of Alexander Fleming alone became inseparably linked with penicillin in the minds of people across the world. Florey, Chain, and the Oxford team were sometimes briefly referred to in newspaper reports, but more often they were not. Sometimes they were wrongly described as participants in Fleming's research.

Fleming himself laughed at the wilder inaccuracies of what he called the "Fleming myth." He began to keep a scrapbook of choice articles cut from the newspapers. Sometimes he would pin these articles on the bulletin board to give his colleagues a laugh.

Top: The containers in this bacteria-identification test kit hold various substances. When a microbe is placed in each container, the reactions that occur tell a bacteriologist exactly which microbe it is.

Final Years

Fleming's happiness in these years was marred only by the growing ill health of his wife, Sareen. At first, she accompanied Fleming on all his speaking tours. By 1948, however, she was too exhausted to continue. Fleming and his son Robert watched helplessly as she became weaker and weaker. She died on October 28, 1949.

Fleming was devastated. She had been his companion, friend, and center of his home life for thirty-four years. Her death left him lonely and isolated. The door of his laboratory, usually open to visitors, was firmly shut. He seemed suddenly much older than his sixty-eight years.

Opposite: Alexander Fleming was carried by cheering students at his election as rector of Edinburgh University in 1952.

Only slowly did Fleming's interest in his work pull him from the misery, and he regained something of his old spirit. A young Greek scientist, Dr. Amalia Voureka, had joined Fleming's laboratory after the war. In the years after the loss of Sareen, Voureka became his valued companion, and she and Fleming were married in 1953. At the age of seventy-four, Fleming still worked in the laboratory and was still able to travel. He continued to do both until the day of his death. He died suddenly, on March 11, 1955, of a heart attack.

The Fleming Legacy

Many doctors believe that penicillin is the greatest single medical advance the world has ever known. Before the 1940s, hospitals were full of people with infections that were raging out of control. After the 1940s, damaged health and death from common infections were essentially things of the past. Penicillin not only cured bodies overwhelmed by bacteria, it gave doctors the power to stop bacteria from taking such a hold in the first place.

The road forward was not without problems. It soon became clear that, although penicillin is generally a safe drug, some people are sensitive to it. These few could have severe, sometimes fatal reactions. And, just as Fleming had predicted, some bacteria developed resistance to penicillin.

In Oxford, Howard Florey went on with his work, pushing back new frontiers in the search for other antibiotics to deal with the penicillin-resistant bacteria. He added a second triumph to his career with antibiotics made from other molds that could combat penicillin-defying bacteria. The new drugs could also be used on people allergic to penicillin.

Many different types of penicillin can now be produced, specially structured to be effective for particular infections, and given by mouth as well as injection. Some combat many more varieties of bacteria than the original drug.

Chance, and the Prepared Mind

Fleming's discovery of penicillin, and of lysozyme, are examples of chance happenings that were turned into scientific riches. Louis Pasteur once said, "In the field of experimentation chance favors the prepared mind." There is no more vivid evidence of this statement than the story of Fleming's scientific gift to the world. The chance arrival of a rare strain of a mold, on a dish of bacteria believed to be invulnerable to attack, was greeted by the trained, perceptive, and curious mind of a scientist who let no random occurrence go by without examining it more closely.

Fleming often said, "I did not invent penicillin. Nature did that. I only discovered it by accident." But his preparedness of mind, sharpened by years in the quest for a perfect antiseptic, prompted him to notice, record, investigate, and preserve a strain of mold almost unique in its ability. It was fifteen years and thousands of mold investigations later before another strain of penicillium with such powers was discovered.

The story of penicillin reflects the many facets of scientific skill and effort by which the world's knowledge advances, pieced together by many people working in different ways and at different times, but all adding to the growth of understanding. Each member of the Oxford team under Florey brought his or her special skills to the disciplined, planned efforts of the group as a whole. All, however, had a central aim fueling their efforts—to develop a substance for safe systemic use in the human body. Fleming, with years of accumulated understanding of bacteria and the destruction they caused, exhibited untiring, wide-ranging, and unbiased investigation of all observations of things that he found interesting. The fascination of the penicillin story lies not only in the knowledge of how the first antibiotic changed the world, but also in understanding the unique and noble qualities of the scientists who produced it successfully.

· ·

"All of us, in our ordinary pursuits, can do research, and valuable research, by continual and critical observation. If something unusual happens, we should think about it and try to find out what it means.... There can be little doubt that the future of humanity depends greatly on the freedom of the researcher to pursue his own line of thought. It is not an unreasonable ambition in a research-worker that he should become famous, but the man who undertakes research with the ultimate aim of wealth or power is in the wrong place."

Alexander Fleming

· ·

Timeline

1860s   Louis Pasteur's work shows that disease is caused by living organisms.

1867   Joseph Lister realizes that bacteria are the cause of infections and uses carbolic acid to clean microbes from wounds and equipment.

1881   Alexander Fleming is born near Darvel, Scotland.

1895   Fleming moves to London.

1897   Fleming begins work as a shipping clerk.

1901   Fleming enters St. Mary's Hospital Medical School.

1906   Fleming passes his final medical school exams. He joins Almroth Wright's department at St. Mary's Hospital.

1909   Fleming passes his surgeon exams, but decides to stay on in Wright's department.

1910   Paul Ehrlich, a German, discovers 606 or salvarsan, which is poisonous to the bacteria which cause syphilis. Fleming becomes well known for his treatment of syphilis.

1914   World War I begins. Almroth Wright's team, including Fleming, goes to Boulogne, France, to treat wounded troops.

1915   Fleming marries Sally McElroy, who is later known as Sareen.

1919   Fleming returns to London.

1921   Fleming and his wife buy their house, The Dhoon, in Suffolk. Fleming discovers lysozyme.

1924   A son, Robert, is born to the Flemings.

1928   Fleming discovers the penicillium mold. He becomes professor of bacteriology in St. Mary's Hospital Medical School.

1929   Fleming names his new discovery "penicillin."

1935   Gerhard Domagk discovers a dye called Prontosil which, when swallowed or injected, kills the streptococci bacteria in mice. This leads to the development of similar substances, known as sulfonamides, to kill certain bacteria.

1940   Ernst Chain extracts enough penicillin to test on animals. Howard Florey successfully injects the penicillin into five mice. Fleming sees the Oxford work on penicillin for the first time.

1941   The Dunn School amasses enough penicillin to plan the first test on a human.

Albert Alexander is treated with the penicillin by the Oxford team for a rosebush scratch. He dies, but further human trials are successful.

1942   Fleming injects penicillin into Harry Lambert's spinal fluid, causing a full recovery. Second human trials by the team in Oxford are successful.

1944   Fleming is knighted for his discovery of penicillin. Howard Florey is knighted for the development of penicillin as a drug.

The first of the new antibiotics after penicillin, streptomycin, is developed.

1945   Fleming, Chain, and Florey are jointly awarded the Nobel Prize for Medicine.

1946   Fleming becomes director of the inoculation department at St. Mary's Hospital when Almroth Wright retires.

1949   Sareen Fleming dies.

1953   Fleming marries Amalia Voureka.

1955   Fleming dies of a heart attack.

Glossary

Abscess: A collection of pus that forms near inflamed tissue.

Antibiotic: A type of medicine made from a microbe that destroys other microbes.

Antiseptic: A chemical substance used to destroy microbes.

Bacillus (plural bacilli): Rod-shaped bacteria.

Bacteria (singular bacterium): One kind of microbe—an organism too small to be seen with the naked eye. The people who study them are bacteriologists.

Coma: A state of unconsciousness from which a person cannot be roused.

Culture: The artificial growth of microorganisms, especially bacteria, in a prepared culture dish.

Epidemic: The spread of a disease to a region in which the disease is not normally present, affecting large numbers of people.

Incubate: To keep something at the right temperature to encourage its growth.

Immune: The state of being resistant to microbes of a particular disease. The body's immune system creates immunity.

Inoculate: To plant a disease in a person or animal by introducing microbes or a virus in order to force the body to develop immunity to it.

Lysozyme: A natural antiseptic, found in human tears, mucus, saliva, pus, and blood serum and in some plants, which fights against infection by microbes.

Meningitis: Inflammation of the membranes that surround the brain and the spinal cord, caused by infection by microbes.

Microorganism: A microscopic (too small to be seen with the naked eye) living creature.

Phagocytes: The white blood cells that swallow and digest bacteria and dead cells.

Staphylococcus (plural staphylococci): Round bacterium that grows in clumps.

Streptococcus (plural streptococci): Round bacterium that grows in chains.

Sulfonamides: The group of chemical drugs that killed certain types of bacteria.

Systemically: Introduced to circulate through the bloodstream.

Vaccine: The slightly weakened bacterium or virus of any disease that is injected into a person or animal in order to create immunity.

Virulent: Extremely dangerous or poisonous.

Virus: The smallest known microbe, much smaller than a bacterium.

For More Information

Books

Deprau, Jeanne. *Cells*. San Diego: Kidhaven, 2001.

Nardo, Don. *Vaccines*. San Diego: Lucent Books, 2001.

Parker, Steve. *Alexander Fleming*. Portsmouth, NH: Heinemann, 2001.

Snedden, Robert. *The Benefits of Bacteria*. Portsmouth, NH: Heinemann, 2000.

Tames, Richard. *Penicillin: A Breakthrough in Medicine*. Portsmouth, NH: Heinemann, 2001.

Tocci, Salvatore. *Alexander Fleming: The Man Who Discovered Penicillin*. Berkeley Heights, NJ: Enslow, 2002.

Websites

All About Vaccines
www.fda.gov/oc/opacom/kids/html/vaccines.htm

How Penicillin Kills Bacteria
www.cellsalive.com/pen.htm

Sir Alexander Fleming Biography
www.nobel.se/medicine/laureates/1945/fleming-bio.html

Index